W9-COO-736

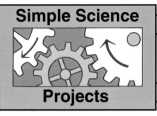

Simple Science Projects

PROJECTS WITH

TIME

By
John Williams

Illustrated by
Malcolm S. Walker

Gareth Stevens Children's Books
MILWAUKEE

For a free color catalog describing Gareth Stevens' list of high-quality books, call 1-800-341-3569 (USA) or 1-800-461-9120 (Canada).

Titles in the Simple Science Projects series:

Simple Science Projects with Air
Simple Science Projects with Color and Light
Simple Science Projects with Electricity
Simple Science Projects with Flight
Simple Science Projects with Machines
Simple Science Projects with Time
Simple Science Projects with Water
Simple Science Projects with Wheels

Library of Congress Cataloging-in-Publication Data

Williams, John.
    Projects with time / John Williams : illustrated by Malcolm S. Walker.
        p.   cm. -- (Simple science projects)
    Rev. ed. of: Time. 1990.
    Includes bibliographical references and index.
    Summary: Explores the concept of time through a variety of projects involving sundials, water clocks,
and other time-keeping devices.
    ISBN 0-8368-0770-7
    1. Time--Juvenile literature. 2. Clocks and watches--Design and construction--Juvenile literature. [1. Time--
Experiments. 2. Clocks and watches--Experiments. 3. Experiments.] I. Walker, Malcolm S., ill. II. Williams,
John. Time. III. Title. IV. Series: Williams, John. Simple science projects.
    QB209.W52 1992
    529'. 076--dc20   91-50548

North American edition first published in MDCCCCLXXXXII by

Gareth Stevens Publishing
1555 North RiverCenter Drive, Suite 201
Milwaukee, Wisconsin   53212, USA

Editor (U.K.): Anna Girling
Editor (U.S.): Eileen Foran
Editorial assistant (U.S.): John D. Rateliff
Designer: Kudos Design Services
Cover design: Sharone Burris

Printed in the United States of America

2 3 4 5 6 7 8 9 9 97 96 95 94

CONTENTS

Words printed in **boldface** type appear in the glossary on pages 30-31.

CLOCKS

Clocks tell us the time. They tell us when to get up in the morning and when to go to bed at night. They tell us when to go to school, or what time to catch a train or bus.

This clock has **Roman numerals**. *What time is it?*

Making a model clock

You will need:

Cardboard
A cardboard box
Scissors
Felt-tip pens
A paper fastener
Glue

1. Cut out a circle from a piece of cardboard. You can draw around a plate to get the right shape.

2. Draw the numbers onto the **dial**. Look at a real clock to see where the numbers should go.

3. Cut two strips of cardboard for the hands of the clock. Make one shorter than the other. Use the paper fastener to attach them to the center of the dial.

4. Attach your dial to the box. Now decorate the front and the sides of the box.

DAY AND NIGHT

Making a "My Day" time line

You will need:

A long piece of poster board
Scissors
Felt-tip pens
A ruler

1. Cut out a long piece of poster board. Use a pen and ruler to divide it into sections.

2. Draw pictures in each section, showing what you do during the day and night. Start with a picture of yourself asleep in bed.

Bed   Breakfast   School   Lunch   More school   Afternoon snack   Play

3. Fold the poster board along each line to make a zigzag book, as shown below.

Making a "My Day" picture clock

You will need:

Poster board
Felt-tip pens
A ruler

1. Draw a circle on a piece of poster board.

2. In your circle, draw little pictures to show what you do at different times during the day — just like you did with the time line.

3. Draw one for different days of the week and one for the weekend. Are they the same?

Have you ever seen a clock like this one in a garden or a public park? The dial is made from all kinds of flowers.

THE SEASONS

You can often tell what **season** it is just by looking out your window. If you have a garden, you can see what flowers are growing there. Some, like daffodils, bloom only in the spring. Others, like poppies, bloom in the summer. When leaves turn yellow or red, it's a sign that fall has arrived. In winter, there may be frost and snow.

It is a bright, sunny day, but what things in this picture tell us it must be fall?

Making a zigzag book of the seasons

You will need:

Poster board
Scissors
Felt-tip pens
A ruler

1. Cut out a strip of poster board and divide it into equal sections to make a zigzag book. You can make one like this one below, to show the different seasons of the year, or you can draw twelve sections — one for each month.

| Winter | Spring | Summer | Fall |

2. Draw pictures in each section to show how the weather, plants, and animals all change with each season of the year.

SUNDIALS

People have been using the Sun for thousands of years to keep proper time. The first sundials were made by the **ancient Egyptians**.

Sundials are easy to make. Here are some different **designs**. For all the following **experiments**, you will need a watch. The watch will tell you when it's time to mark off each hour. You may have to wait for a sunny day!

WARNING: Never look at the Sun directly, even through dark glasses. It can damage your eyes.

This is a simple sundial, like the ones you can make. Look around for sundials on buildings and in parks.

Making a large sundial for your playground or yard

1. Drive a stake or a pole upright into a level piece of cleared ground.

2. Check to see where the shadow is at different times of the day.

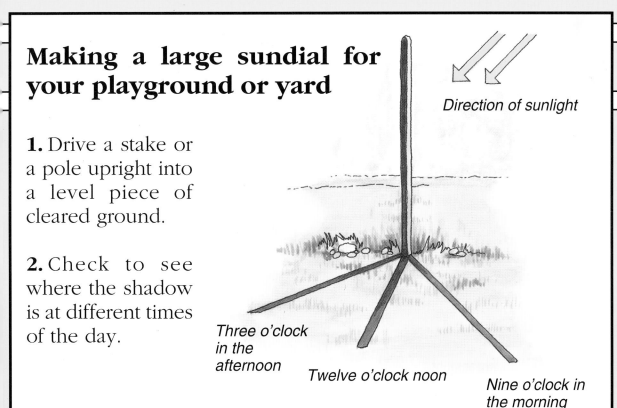

Direction of sunlight

Three o'clock in the afternoon

Twelve o'clock noon

Nine o'clock in the morning

Making a hand-held sundial

You will need:

A piece of wood
A nail
A small hammer
A pencil

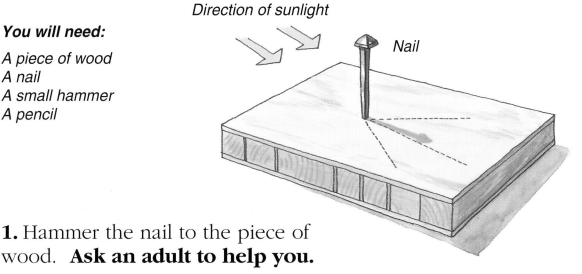

Direction of sunlight

Nail

1. Hammer the nail to the piece of wood. **Ask an adult to help you.**

2. Put your sundial in the sunlight and mark on it where the shadow is at different times of the day. Always keep it facing the same direction.

Making a cardboard sundial

You will need:

Two pieces of cardboard
Scissors
Glue

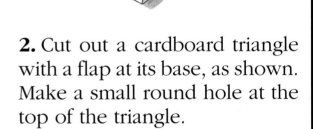

Sun

Shadow

1. Cut out a square piece of cardboard. Fold it along the middle.

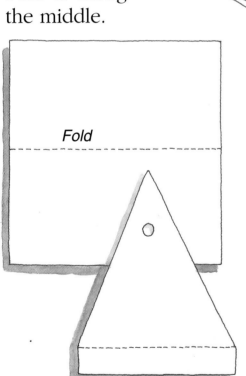

Fold

2. Cut out a cardboard triangle with a flap at its base, as shown. Make a small round hole at the top of the triangle.

3. Glue the flap of the triangle along the fold line of the square piece of cardboard.

4. Put the sundial outside and see how the triangle's shadow moves during the day.

Further work with sundials

Place a small pole in your backyard and mark the shadow as the Sun moves during the day. **Measure** the length of the shadow at different times. When is it longest? When is it shortest?

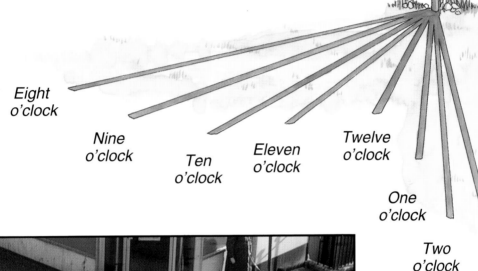

Eight o'clock

Nine o'clock

Ten o'clock

Eleven o'clock

Twelve o'clock

One o'clock

Two o'clock

Three o'clock

On a sunny day, shadows are very noticeable. These shadows are also very long. Can you guess what time of day this might be?

This is a modern water clock. The weight of the water in the side channels makes the pendulum swing back and forth.

Water clocks were first used by the ancient Egyptians 3,500 years ago. They used them to keep time during the night, when they could not use the Sun to tell time. The clocks were made from large bowls. The bowls were filled with water at sunset, and the water trickled out slowly during the night.

Making a water clock

You will need:

Four small plastic cups
A small plastic pail
A wooden stand
Pushpins
A pitcher of water
A timer

1. Use the pushpins to make a very small hole in the bottom of each of the smaller cups.

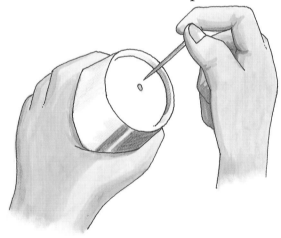

2. Use the pushpins to attach the cups to the stand, as shown. Attach the pail to the bottom of the stand.

3. Fill the top cup with water. Measure the time it takes for the water to trickle down through the cups and into the pail.

OLD WATER CLOCKS

Making an Egyptian-style water clock

You will need:

A large plastic cup
A big bowl
A nail
A pen or marker
A timer

1. Use the nail to make a hole in the bottom of the cup.

2. Put your finger over the hole. Fill the cup with water. Take your finger away and let the water drip out into the bowl. Time how long it takes for the cup to empty.

3. Try the same experiment again. This time, mark the water level on the side of the cup once every minute to make a **scale**. Now you can use it as a kind of clock.

Chinese water clocks

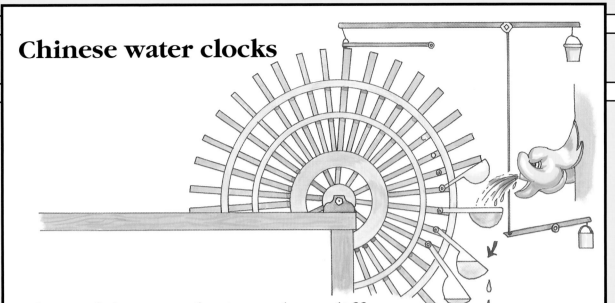

The **Chinese** designed a different kind of water clock. They used a slowly turning wheel like the one shown above. Running water turns the wheel. The wheel is attached to a pointer that shows the time.

This very old water clock is one that was used by the ancient Egyptians. The scale for measuring the water is inside.

CANDLE CLOCKS

A long time ago, candles were used to measure time during the night. The candles were marked into sections of one hour each. As the candle burned down, people could see how much time had passed.

The time scale on this clock is marked in Roman numerals. VI means six and XII means twelve.

Making a candle clock

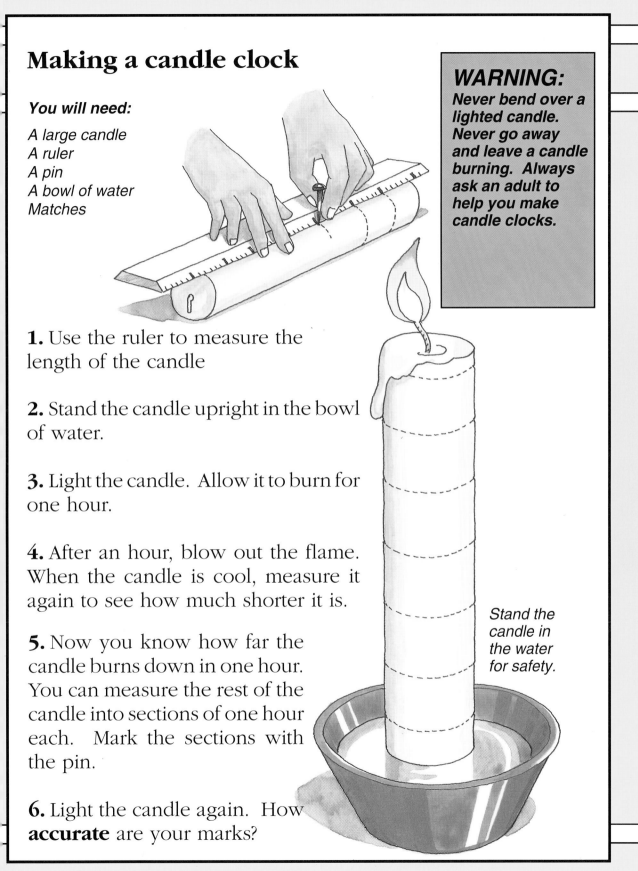

You will need:

A large candle
A ruler
A pin
A bowl of water
Matches

1. Use the ruler to measure the length of the candle

2. Stand the candle upright in the bowl of water.

3. Light the candle. Allow it to burn for one hour.

4. After an hour, blow out the flame. When the candle is cool, measure it again to see how much shorter it is.

5. Now you know how far the candle burns down in one hour. You can measure the rest of the candle into sections of one hour each. Mark the sections with the pin.

6. Light the candle again. How **accurate** are your marks?

Stand the candle in the water for safety.

TIMERS

Sand timers have been used for hundreds of years. They are still used today. You might have one in your kitchen for measuring the time it takes to boil an egg.

Making a sand timer

You will need:

An empty plastic bottle and
 a cap with a hole in it
A glass jar
Scissors
Sand or salt
A timer

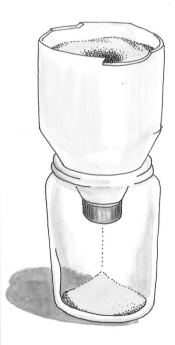

1. Make sure the bottle will fit upside down in the neck of the jar.

2. Use the scissors to cut off the bottom of the bottle.

3. Put the bottle cap back on. Balance the bottle upside down in the jar.

4. Fill the bottle with sand or salt. Time how long it takes to empty into the jar.

Making a marble timer

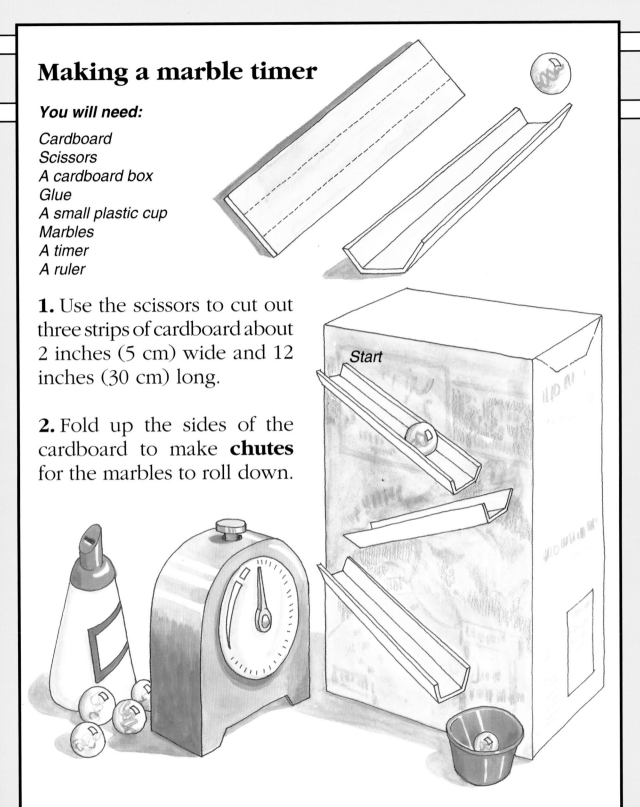

You will need:

Cardboard
Scissors
A cardboard box
Glue
A small plastic cup
Marbles
A timer
A ruler

1. Use the scissors to cut out three strips of cardboard about 2 inches (5 cm) wide and 12 inches (30 cm) long.

2. Fold up the sides of the cardboard to make **chutes** for the marbles to roll down.

Start

3. Glue the cardboard chutes to the box, as shown. Try out your timer and see how it works.

PENDULUMS

A lump of modeling clay on the end of a piece of string makes a simple pendulum. Pendulums are used as timers in some clocks. They are very accurate timers.

You will need:

Modeling clay
String
Scissors
A long stick
A timer
A ruler

Experimenting with pendulums

1. Make a pendulum 2 feet (60 cm) long by attaching a ball of modeling clay to a piece of string.

2. Tie the stick onto the backs of two chairs. Hang your pendulum from the stick. Make sure it doesn't hit the floor when it swings.

3. Time how many seconds your pendulum takes to swing from side to side thirty times.

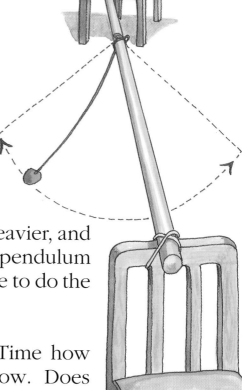

4. Add more clay to make the ball heavier, and try the experiment again. Does the pendulum always take the same amount of time to do the same number of swings?

5. Make your pendulum shorter. Time how long it takes to swing thirty times now. Does it take less time?

Further work

A pendulum swings from side to side. Can you think of objects other than a pendulum that can be used as timers? Try to find some. Here are some ideas.

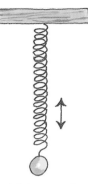

A bobbing spring

A dripping faucet

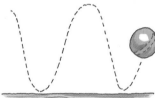

A bouncing ball

A large clock like this needs a long, heavy pendulum. This is Big Ben in London, England. Its pendulum is 13 feet (4 m) long.

You might have a **digital** watch of your own. It does not have hands or a clock face. It has numbers that may look like this: 01:00 or 08:30.

Here you can see a twenty-four hour digital clock at a train station. It shows the hours, minutes, and seconds. Can you figure out what time it would be on a twelve-hour clock? See page 25.

24

Making a twenty-four-hour time chart

You will need:

A large piece of poster board
A ruler
Felt-tip pens

1. Divide the poster board into twenty-four sections, using the ruler and some felt-tip pens. Across the top, number the sections 1-24. Across the bottom, number the sections 1-12 noon (**A.M.**) and 1-12 midnight (**P.M.**).

DIGITAL TWENTY-FOUR HOUR TIME

2. Now you have a chart you can use to translate twenty-four-hour time into regular time. What time is it when it's 07:00 hours? 15:00 hours? 21:00 hours? Draw pictures on your chart to show what you might be doing at these hours.

ALARM CLOCKS

We use alarm clocks for many things. Alarm clocks wake us in the morning. They also tell us when food is done. Think of other times when an alarm clock is used.

You will need:

A candle
Matches
A bowl of water
A pin
String
A metal nut
A spool
A metal tray

Making a candle alarm clock

1. Stand the candle upright in a bowl of water.

2. Tie one end of the string to the nut and the other end to the pin. Stick the pin securely into the candle. **Make sure the nut won't pull the candle over.**

3. Make a simple **pulley** out of the spool and hang the string over it. Attach the pulley to a secure object.

4. Place the tray under the nut.

5. Light the candle. When the candle burns down far enough, the pin falls out. The nut hits the tray, setting off the "alarm."

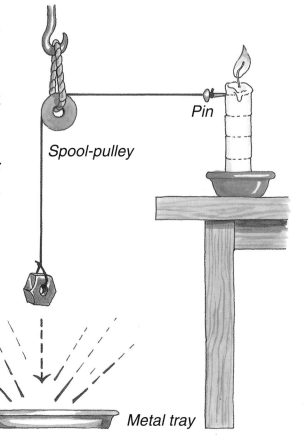

Pin

Spool-pulley

Metal tray

Further work

Make an electric light for your alarm clock.

You will need:

A piece of cardboard
Three pieces of insulated copper wire
Paper fasteners
*A 6V **battery***
A 6V light bulb and bulb holder

1. Fold the cardboard in half.

2. Stick a paper fastener in each half so that the heads touch when you close the cardboard.

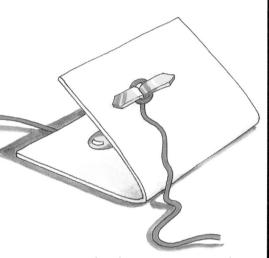

3. Attach the wires to the outside of each fastener.

4. Connect the battery, paper fasteners, and bulb, as shown.

5. Try the experiment on page 26 again. This time, the metal nut should fall on the cardboard, rather than on the metal tray. The falling nut will close the cardboard and light the bulb. You have made an **electric circuit**.

Falling metal nut closes the circuit

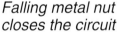

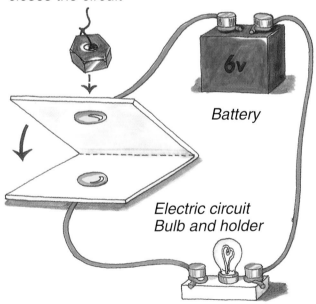

Battery

Electric circuit Bulb and holder

What You'll Need

More Books About Time

All of Grandmother's Clocks: A Beginning Book About Time. Sandra Ziegler
  (Child's World)
Anno's Sundial. Mitsumasa Anno (Putnam)
Clocks and How They Go. Gail Gibbons (Harper & Row Junior Books)
Clocks and Time. Ed Catherall (Silver, Burdett, & Ginn)
How Did We Get Clocks and Calendars? Susan Perry (Creative Education)
The Secret Clocks: Time Senses of Living Things. Seymour Smith (Penguin)
Time! Jane Edmonds and Mark Sachner (Gareth Stevens)
Time and Clocks. Herta S. Breiter (Raintree)
What Time Is It Around the World? Hans Baumann (Scroll)

More Books With Projects

Clocks: Building and Experimenting with Model Timepieces. Bernie Zubrowski
  (Morrow Junior Books)
Einstein Anderson Makes Up for Lost Time. Seymour Simon (Viking Penguin)
Make It with Odds and Ends. Felicia Law (Gareth Stevens)
Shadow Magic. Seymour Simon (Lothrop, Lee, & Shepard)
*Sun Dials and Time Dials: A Collection of Working Models to Cut and Glue
  Together.* Gerald Jenkins and Magdalen Bear (Tarquin)

Places to Write for Science Supply Catalogs

The Nature of Things
275 West Wisconsin Avenue
Milwaukee, Wisconsin 53203

Suitcase Science
Small World Toys
P. O. Box 5291
Beverly Hills, California 90209

Nasco Science
901 Janesville Road
Fort Atkinson, Wisconsin 53538

Adventures in Science
Educational Insights
19560 Rancho Way
Dominguez Hills, California 90220

Ward's Natural Science
P. O. Box 1712
Rochester, New York 14603

Schoolmasters Science
P. O. Box 1941
Ann Arbor, Michigan 4810

GLOSSARY

accurate
Exact and correct.

A.M.
Before noon; in the morning.

ancient Egyptians
The people who lived in Egypt 2,000 to 5,000 years ago. The Egyptians invented many of the earliest ways to measure time.

battery
A device that makes or stores electricity.

Chinese
The people who live in China. The Chinese invented many things we use today, including paper, printing, gunpowder, and the compass. Chinese history dates back about 3,000 years.

chute
A shallow, gutter-shaped trough or channel.

designs
Drawings or patterns used for planning and making an object or a machine.

dial
A circle with numbers around it. The face on a clock is one kind of dial. The knob on a radio that chooses stations is another.

digital
Using numbers instead of hands to show the time on a clock.

electric circuit
A loop of wires and objects connected so that electricity will flow through it.

experiment
A test to find out whether or not an idea works.

measure
To find out how big or heavy an object is or how long something takes to happen.

P.M.
After noon; in the afternoon or evening.

pulley
A wheel with a rope around it. Pulleys are used to make heavy objects easier to lift.

Roman numerals
An old system of counting that uses I for 1, V for 5, and X for 10. In Roman numerals, "XII" equals 12 (10 + 1 + 1).

scale
A line of regular marks used for measuring.

season
One of the four parts of the year. Spring, summer, fall, and winter are the seasons.

Picture acknowledgements
The publishers would like to thank the following for allowing their photographs to be reproduced in this book: Cephas Picture Library, p. 8; Chapel Studios, p. 24; Eye Ubiquitous, pp. 4, 13; Michael Holford, p. 17; Hutchison Library (J.G.Fuller), p. 7; PHOTRI, p. 10; The Time Museum, Rockford, Illinois, p. 14; Tim Woodcock, p. 23; Zefa Picture Library, p. 18. Cover photography by Zul Mukhida.

INDEX